FOCUS ON
ELEMENTARY
PHYSICS

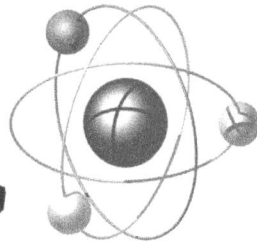

Laboratory Notebook
3rd Edition

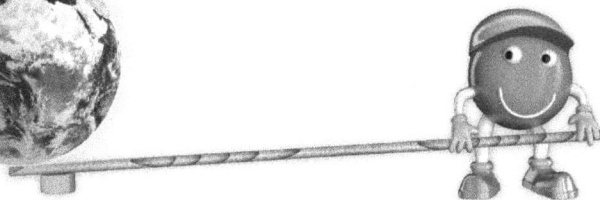

Rebecca W. Keller, PhD

REAL SCIENCE 4 Kids

Real Science-4-Kids

Illustrations: Janet Moneymaker

Focus On Elementary Physics Laboratory Notebook—3rd Edition
ISBN 978-1-941181-43-0

Published by Gravitas Publications Inc.
www.gravitaspublications.com
www.realscience4kids.com

GRAVITAS
PUBLICATIONS

A Note From the Author

Hi!

In this curriculum you are going to learn the first step of the scientific method:

Making good observations!

In the science of physics, making good observations is very important.

Each experiment in this notebook has several different sections. In the section called *Observe It*, you will be asked to make observations. In the *Think About It* section you will answer questions. There is a section called *What Did You Discover?* where you will write down or draw what you observed from the experiment. And finally, in the section *Why?* you will learn about the reasons why you may have observed certain things during your experiment.

These experiments will help you learn the first step of the scientific method and.....they're lots of fun!

Enjoy!

Rebecca W. Keller, PhD

Contents

Experiment 1

Falling Objects

I. Observe It

❶ Take two tennis balls and hold them at chest level with your arms pointing straight out in front of you.

❷ Release the two objects from both hands at the same time.

❸ Watch carefully to see how they land.

❹ In the following box use words or pictures to record what you see.

❺ Repeat Steps ❶-❹ several times using different objects, such as:

- ◆ An orange and an apple.

- ◆ A tennis ball and a rubber ball.

- ◆ An apple and a tennis ball.

- ◆ A rubber ball and an apple.

- ◆ An orange and a tennis ball.

Object 1 _____

Object 2 _____

_____

_____

_____

_____

_____

_____

Object 1 _____

Object 2 _____

_____

_____

_____

_____

_____

_____

_____

Object 1 _____

Object 2 _____

_____

_____

_____

_____

_____

_____

Object 1 _____

Object 2 _____

_____

_____

_____

_____

_____

_____

_____

Object 1 _____

Object 2 _____

_____

_____

_____

_____

_____

_____

_____

Object 1 _____

Object 2 _____

_____

_____

_____

_____

_____

_____

_____

Object 1 _____

Object 2 _____

_____

_____

_____

_____

_____

_____

_____

II. Think About It

❶ Did the objects fall at the same speed? How can you tell?

_____

_____

_____

_____

_____

❷ Are there any changes you could make to your experiment? Holding the objects higher? Holding the objects lower? Describe changes you can make.

_____

_____

_____

_____

_____

_____

_____

_____

❸ Repeat the experiment for one set of objects using one change you thought about.

❹ Record your observations on the next page.

Object 1 _____

Object 2 _____

Change to Experiment _____

_____

_____

_____

_____

_____

_____

_____

_____

III. What Did You Discover?

❶ Was it easy or difficult to release the objects at the same time? Why or why not?

_____

_____

_____

❷ Was it easy or difficult to observe the objects falling? Why or why not?

_____

_____

_____

❸ For each of the pairs of objects, did both objects land at the same time? Why or why not?

_____

_____

_____

❹ Did the changes you chose to make to your experiment make a difference? Why or why not?

_____

_____

_____

_____

_____

IV. Why?

Galileo Galilei discovered that when he let two objects of different weights fall from the same height, they always landed at the same time. This seems the opposite of what you might think would happen. It seems like a heavier object would fall faster than a lighter object. But this is not what happens. Your observations showed that two objects of different weights will hit the ground at the same time. Why?

Things fall because of gravity. Gravity is a force that makes the objects on Earth stay on Earth. (You will learn about forces in a following chapter.) Gravity pulls everything down towards the center of the Earth. When you hold two objects in your hands, gravity is pulling on them. Every object has gravity pulling on it all the time. Gravity pulls on apples in the same way that it pulls on tennis balls. Gravity pulls on oranges in the same way that it pulls on rubber balls. Everything has the same force of gravity pulling on it at the same time. So an apple (which is heavier than a tennis ball) has the same amount of gravity pulling on it as the tennis ball. Both the tennis ball and the apple start off with exactly the same amount of gravity pulling on them at the same time, and the amount of gravity pulling on them never changes.

Once the objects are released, they fall at the same speed because they have the same amount of gravity pulling on them at the same time. It doesn't matter how heavy they are. That is what Galileo and YOU discovered by doing this experiment.

V. Just For Fun

What do you think would happen if you dropped an orange and a cotton ball or a feather at the same time? Why?

Try it. Record your observations.

Object 1   *Orange*

Object 2   *Cotton Ball (or Feather)*

Experiment 2

Measuring Time

Introduction

One important measurement physicists make is time. How long do events take? In this experiment you will use a timer to measure how long certain events take.

I. Think About It

How long does it take for a marble to roll down a ramp? How long does it take for a car to travel from point A to point B? How long does it take for the Sun to rise or set? How much time is there between a high tide and a low tide? Think of three events you would like to measure and write or draw them below.

3 Events to Measure

II. Observe It

For each of the three events you listed in the *Think About It* section, use a timer to measure how long it takes. Then repeat and time it again. You will be measuring each event twice. Record the times and any observations in the following boxes.

Event 1 _____

Event Times and Observations

Event 2_____

Event Times and Observations

Event 3 _____

Event Times and Observations

III. What Did You Discover?

❶ How easy (or difficult) was it to measure time? Why?

_____

_____

❷ Did each event take the same amount of time both times? Why or why not?

_____

_____

❸ What was the longest time you recorded?

_____

_____

❹ What was the shortest time you recorded?

_____

_____

❺ How easy (or difficult) was it for you to record the times of these events? Why?

_____

_____

❻ If you didn't have a timer, how would you measure time?

_____

_____

_____

IV. Why?

Time is an important part of understanding how things work. Knowing how long something takes gives physicists information about how to calculate things such as speed or how long something needs to be kept at a specific temperature to get a certain result. For example, a physicist can figure out how fast a train was going by knowing how far the train traveled and how long it took for the train to arrive. This is one way to measure speed. And when you are baking a cake, you need to know the amount of time it takes to bake it at a specific temperature.

V. Just For Fun

How fast can you run? Pick a distance you can run. Take your timer and measure how long it takes you to run that distance. Repeat until you are out of breath. Did you run faster or slower the first time? The other times? In the space below and on the next page, record your times and anything else you observe.

Running—Time Measurements and Observations

Running—Time Measurements and Observations

Experiment 3

Get To Work!

I. Observe It

❶ Take a marshmallow and observe its color, shape, and size. Write or draw your observations in the "Before" space on the following page.

❷ Take the marshmallow and place it in the center of your palm.

❸ Close your hand around the marshmallow and squeeze it with your palm and fingers.

❹ Observe your effort—muscles, hands, and fingers.

❺ Observe the marshmallow after you have squeezed it. Write or draw your observations in the space next to "After."

❻ Repeat Steps ❶-❺ with several other objects such as:

- ◆ rubber ball
- ◆ tennis ball
- ◆ lemon or lime
- ◆ rock
- ◆ banana

Marshmallow

Before

_____
_____
_____
_____
_____
_____
_____
_____
_____
_____

After

_____
_____
_____
_____
_____
_____
_____
_____
_____
_____

_____

Before

_____

_____

_____

_____

_____

_____

_____

_____

_____

_____

After

_____

_____

_____

_____

_____

_____

_____

_____

_____

_____

Before

_____
_____
_____
_____
_____
_____
_____
_____
_____
_____

After

_____
_____
_____
_____
_____
_____
_____
_____
_____
_____

_____

Before

_____
_____
_____
_____
_____
_____
_____
_____
_____
_____

After

_____
_____
_____
_____
_____
_____
_____
_____
_____

Before

_____

_____

_____

_____

_____

_____

_____

_____

_____

_____

After

_____

_____

_____

_____

_____

_____

_____

_____

_____

_____

_____

Before

_____

_____

_____

_____

_____

_____

_____

_____

_____

_____

After

_____

_____

_____

_____

_____

_____

_____

_____

_____

_____

Before

_____

_____

_____

_____

_____

_____

_____

_____

_____

_____

After

_____

_____

_____

_____

_____

_____

_____

_____

_____

_____

II. Think About It

❶ How did the objects feel in your hands?

_____

_____

_____

_____

_____

❷ Were some objects easier to squeeze in your hand than other objects? Were some objects harder to squeeze in your hand than other objects?

_____

_____

_____

_____

_____

❸ On the next page, create a summary of your observations. List those objects that were easy to squeeze and those objects that were hard to squeeze.

❹ Put a circle around the object on which you believe your hand did the most work.

❺ Put a rectangle around the object on which you believe your hand did the least work.

Easy to Squeeze	Hard to Squeeze
_____	_____
_____	_____
_____	_____
_____	_____
_____	_____
_____	_____
_____	_____
_____	_____
_____	_____
_____	_____

III. What Did You Discover?

❶ For the objects that were easy to squeeze, how much force did you need?

_____

_____

❷ For the objects that were hard to squeeze, how much force did you need?

_____

_____

❸ Did you do more or less work on the objects that were easy to squeeze?

_____

_____

❹ Did you do more or less work on the objects that were hard to squeeze?

_____

_____

❺ Was this result what you expected? Why or why not?

_____

_____

_____

_____

_____

IV. Why?

In this experiment you squeezed several objects between the palm and fingers of your hand. You used your hand like a little force tool. You can do this because your hand has lots of nerve endings that can detect how soft or hard an object is. You can also tell with your body how easy or hard it is to squeeze an object by noticing your muscles and breath. You can tell with your body if more or less force is needed to change the shape of an object. You can also tell with your body if more or less energy is required to generate the force.

Do you do more work if you use more force and more energy? Not necessarily. You may have noticed that you could easily smash the marshmallow, but it was harder to smash the rubber ball, tennis ball, or rock. Depending on your results, you may have used more force to try to change the shape of the tennis ball or rock, but if they did not actually change shape, you didn't end up doing any work!

V. Just For Fun

First, squeeze a tennis ball with your hand. Then, use a pair of pliers to squeeze the tennis ball. Observe whether the tennis ball changes shape. Also observe whether it is easier or harder to change the shape of the tennis ball with a pair of pliers than with your hand.

Record your observations.

Tennis Ball

Without Pliers	With Pliers
_____	_____
_____	_____
_____	_____
_____	_____

Experiment 4

Moving Energy in a Toy Car

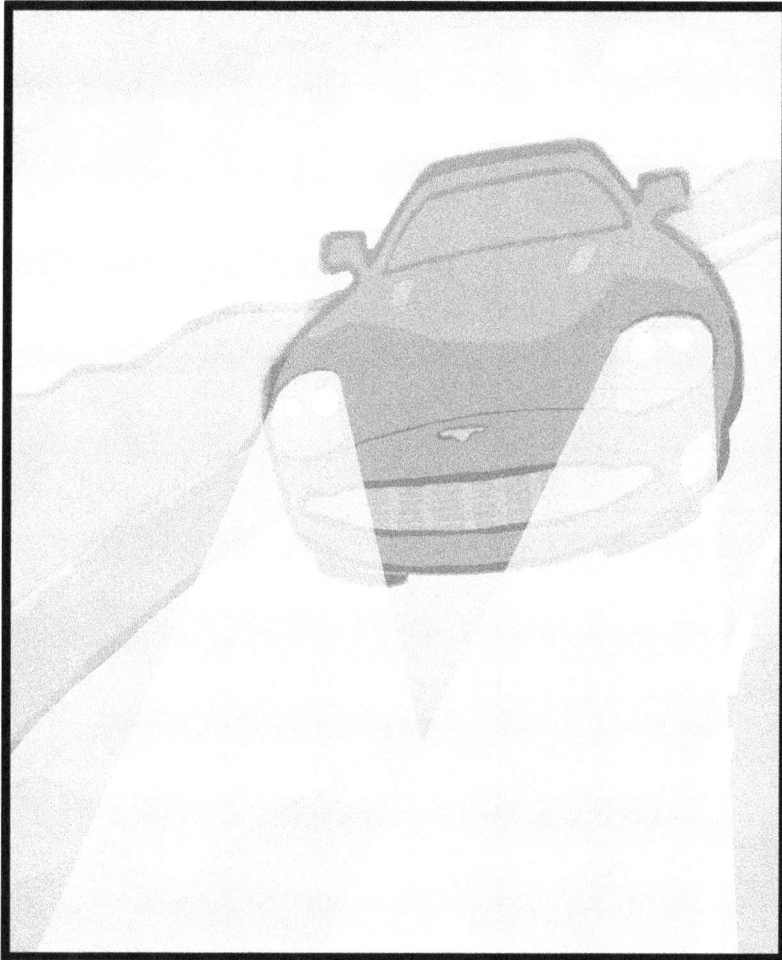

I. Observe It

❶ Take the board or sheet of cardboard and lay it flat on the ground. Place the toy car on the board or cardboard sheet. Without rolling the toy car, observe what happens.

❷ Write or draw your observations in the space below.

Car — flat

❸ Now, lift the board or cardboard sheet to the height of your ankles, making a ramp. Again observe what happens. Note whether or not the toy car moves. If it does move, note how it moves.

❹ Write or draw your observations in the space below.

Car lifted to ankles

_____

_____

_____

_____

_____

_____

_____

_____

_____

_____

_____

_____

_____

_____

_____

_____

_____

❺ Now lift the ramp to the height of your knees. Again observe what happens. Note whether or not the toy car moves. If it does move, note how it moves.

❻ Write or draw your observations in the space provided below.

Car lifted to knees

_____

_____

_____

_____

_____

_____

_____

_____

_____

_____

_____

_____

_____

_____

_____

_____

❼ Now lift the ramp to the height of your hips. Again observe what happens. Note whether or not the toy car moves. If it does move, note how it moves.

❽ Write or draw your observations in the space provided below.

Car lifted to hips

_____

_____

_____

_____

_____

_____

_____

_____

_____

_____

_____

_____

_____

_____

_____

_____

❾ Now lift the ramp to the height of your chest. Again observe what happens. Note whether or not the toy car moves. If it does move, note how it moves.

❿ Write or draw your observations in the space provided below.

Car lifted to chest

_____

_____

_____

_____

_____

_____

_____

_____

_____

_____

_____

_____

_____

_____

Collect Your Results

In the space below create a table showing your results.

Height of Car	Observations
Car flat	
Car lifted to ankles	
Car lifted to knees	
Car lifted to hips	
Car lifted to chest	

II. Think About It

❶ Think about what the toy car did as you lifted the ramp higher and higher.

❷ Review what you learned in your *Student Textbook* about gravitational stored energy. Recall that gravitational stored energy exists in objects that are elevated above the ground.

❸ Think about both the toy car and gravitational stored energy. Guess which car height had the least amount of gravitational stored energy and which car height had the greatest amount of gravitational stored energy. Record your answers.

Least _____

Most _____

❹ In the chart on the following page, put a circle around the car that had the least amount of gravitational stored energy.

❺ Put a rectangle around the car that had the greatest amount of gravitational stored energy.

Car flat

Car lifted to ankles

Car lifted to knees

Car lifted to hips

Car lifted to chest

III. What Did You Discover?

❶ How high did you need to lift the board or cardboard sheet before the car began to move?

_____

_____

❷ Why do you think the car on the ground did not move?

_____

_____

❸ Do you think that as you lifted the ramp higher and higher, the car gained more and more gravitational stored energy? Why or why not?

_____

_____

❹ Do you think there was more kinetic energy (moving energy) in the car as you lifted the ramp? Why or why not?

_____

_____

_____

❺ Was this result what you expected? Why or why not?

_____

_____

_____

IV. Why?

In this experiment you took a toy car, and placing it on a board or cardboard sheet, you observed how the amount of gravitational stored energy changed as you lifted the board. In the first observation, you saw that the car did not move. When the car was not elevated and was sitting flat on the ground, the car did not have any gravitational stored energy. As you lifted the car to your ankles, the car was "given" gravitational stored energy by your body. The car may not have moved until you lifted the ramp to your knees or to your hips, but each time you lifted the car you "gave" it more gravitational stored energy until there was enough stored energy for the car to move down the ramp.

When the car moved down the ramp, the gravitational stored energy was converted to kinetic (or moving) energy. You may have observed that the higher you lifted the toy car, the faster the car moved down the ramp. Since the toy car has more gravitational stored energy as it is lifted higher, there is more energy to convert to kinetic energy. As a result, there is more kinetic energy as the car goes down the ramp.

V. Just For Fun

How high do you have to lift the toy car to smash a marshmallow placed at the end of the ramp?

Put a few marshmallows at the end of your ramp. Raise the ramp, and let the toy car roll down. See if you can get the toy car to hit a marshmallow. How high do you have to lift the car for it to smash the marshmallow? Record your observations.

Height of Car	Did it smash the marshmallow?	
Car flat	YES	NO
Car lifted to ankles	YES	NO
Car lifted to knees	YES	NO
Car lifted to hips	YES	NO
Car lifted to chest	YES	NO

Hint: You may have to lift the ramp to your head or higher! Also, what would happen if you used a heavier car?

One way scientists make discoveries is by changing a part of their experiment to see what will happen. Try making other changes to this experiment. What would happen if you used a longer ramp? A shorter ramp? What other changes could you make? Try one or more of them and see what you discover.

Changes to the Experiment
and Results of the Changes

Experiment 5

Playing With Physics

I. Observe It

❶ Take two marbles. Roll one marble into the other marble.

❷ Draw what happens to the two marbles.

Two Marbles

❸ Take three playing cards and make a small card house.

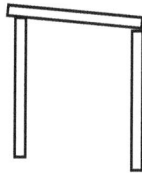

❹ Roll a marble so it hits the card house. Draw what happens to the cards.

Marble Hits Card House

❺ Find a shallow jar top. A pickle jar top would work well. Fill it with vinegar.

❻ Add 15 milliliters (1 tablespoon) of baking soda to the vinegar.

❼ Draw what happens.

Vinegar and Baking Soda

❽ Fill the shallow lid with vinegar again, and make a card house above the lid.

❾ Place 15 milliliters (1 tablespoon) of baking soda on top of the card house. Tip the card house with your finger so the baking soda falls into the vinegar. Record your observations below.

Card House Plus Baking Soda and Vinegar

⑩ Now assemble all the steps into a short series. Take one marble and place it a few inches from the card house. Place the shallow lid of vinegar under the card house, then place a tablespoon of baking soda on top of the card house. Roll a second marble into the marble that is close to the card house. Record your observations below.

Marble-->Marble-->Card House w/Baking Soda-->Vinegar

II. Think About It

❶ Think about the different types of energy you assembled in this short series.

❷ Review what you learned about energy in this chapter of the *Student Textbook*. Note how energy is neither created nor destroyed but simply converted from one form to another.

❸ In the table below list the type of energy you think the object started with and the type of energy you think it was converted to. Use the following energy descriptions:

> Kinetic energy (rolling)
> Kinetic energy (falling)
> Stored energy (chemical)
> Stored energy (gravitational)
> Chemical energy

Object	Started With	Converted To
marble-->marble		
marble-->card house		
card house upright--> card house falling		
baking soda or vinegar -->baking soda + vinegar		

III. What Did You Discover?

❶ What happened to the energy of one marble when you rolled it into another marble that was not moving?

_____

_____

❷ What happened to the energy of the marble when you rolled it and it hit the card house?

_____

_____

_____

❸ What happened to the energy in the baking soda (or the vinegar) when you added the two together?

_____

_____

_____

❹ What happened to the energy of the marble you rolled when you put all the steps together?

_____

_____

_____

❺ What do you think happened to all the energy at the end of your experiment? Where did it go?

_____

_____

_____

_____

IV. Why?

In this experiment you explored how different forms of energy can be converted from one form to another. You rolled a marble and watched it use kinetic energy to move. You had that marble strike another marble, and you observed how the kinetic energy of the first marble was converted into kinetic energy in the second marble.

You also rolled a marble to hit a card house, converting the kinetic energy of the rolling marble into the kinetic energy of the falling cards. You observed how the gravitational stored energy of the top card was converted into kinetic energy when it fell.

When you put the vinegar underneath the card house and placed baking soda on top, you converted chemical stored energy into chemical energy when the baking soda and vinegar came together.

In each case, you observed energy being converted from one form into another. This is how energy works. You can't create energy, and you can't destroy it. You can only move it from one object to another or convert it from one form to another.

What do you think happened to all the energy at the end of your experiment? Where did it go?

V. Just For Fun

Create your own series for converting energy from one form to another.

Here are some ideas—see if you can connect them.

- Rolling marbles

- Dominoes side by side

- Vinegar and baking soda

- Stacked blocks

- An electric car

- An electric train

- Marshmallow on one end of a tongue depressor, a steel ball or marble dropped on the other end

Series Ideas

_____

_____

_____

_____

_____

_____

_____

_____

_____

_____

Draw your series setup.

Write or draw what happened.

_____

_____

_____

_____

_____

_____

_____

_____

_____

_____

Experiment 6

Rolling Marbles

Introduction

In this experiment you will use marbles to explore the effects of inertia, mass, and friction.

I. Observe It

❶ Take two marbles—one small glass marble and one large glass marble. Roll them both across a smooth surface (wooden floor or tabletop).

❷ Observe what happens. How far do they go? Do they go straight? How do they stop?

❸ Write or draw a description below of how the marbles rolled. Include any details you think might be important.

Small and Large Marbles Rolling on Smooth Surface

_____

_____

_____

_____

❹ Take two marbles—one small glass marble and one large glass marble—and roll them across a rough surface (carpeted floor or grass lawn).

❺ Observe what happens. How far do they go? Do they go straight? How do they stop?

❻ Write or draw a description below of how the marbles rolled. Include any details you think might be important.

Small and Large Marbles Rolling on Rough Surface

_____

_____

_____

_____

II. Think About It

❶ Review what you learned in the *Student Textbook* about mass, inertia, and friction.

❷ Thinking about the two surfaces (smooth and rough), put a circle around the surface that has the most friction.

Rough Surface

Smooth Surface

❸ Thinking about the two marbles (large and small) put a box around the size that has the most mass.

Large Marble

Small Marble

❹ Thinking about the two marbles (large and small), put a triangle around the size that has the most inertia.

Large Marble

Small Marble

III. What Did You Discover?

❶ Compare how the small marble moved across the smooth surface with how it moved across the rough surface. Did the small marble move differently on different surfaces? Why or why not?

_____

_____

_____

_____

_____

❷ Compare how the large marble moved across the smooth surface with how it moved across the rough surface. Did the large marble move differently on different surfaces? Why or why not?

_____

_____

_____

_____

_____

IV. Why?

In this experiment you observed how two marbles of different size (and mass) moved on both a smooth surface and a rough surface. If you compare your results, you can see how mass and inertia are related and how the force of friction changes how a marble moves. You may have observed that the small marble moved easily on the smooth surface but did not move as easily on the rough surface. You also may have observed that the large marble also moved easily on the smooth surface but not as easily on the rough surface. And you might have observed that the larger marble moved farther on the rough surface than the smaller marble did.

The larger marble has more mass than the smaller marble. Because it has more mass, it has more inertia. Having more inertia means that more force is required to start the large marble rolling. But once the large marble is rolling, more force is also required to make it stop rolling. The same amount of friction will more easily stop a small marble than a large marble because a small marble has less mass.

In your experiment you observed how the force of friction can stop a moving object. What if you could roll marbles in space? Do you think the marbles would stop? No! The marbles would never stop unless they hit something else. Why? Because once the marbles are rolling, only a force can stop them. In space there is no air, and so there is no force of friction to stop them from moving.

V. Just For Fun

What happens if a small marble bumps into a large marble?

Take the large marble and place it in the middle of the table or on the floor. Take the small marble in your fingers and with your thumb shoot the small marble toward the large marble. See if you can hit the large marble with the small marble.

Record your observations on the following page.

Try doing the opposite. Place the small marble in the center of the table and hit it with the large marble.

Record your observations on the following page. Did you notice anything different happening?

Small Marble Hitting Large Marble

_____

_____

_____

_____

Large Marble Hitting Small Marble

_____

_____

_____

_____

Experiment 7

Speed It Up!

Introduction

Understanding speed is an important concept in physics. In this experiment you will explore how to measure speed.

I. Think About It

How long does it take for you to run from one end of your yard (or park) to the other? If it is a short distance, can you run fast? If it is a longer distance, does it take longer? Do you think you can run at the same speed the entire distance? Or do you think you'll have to run fast and then slow down to make it the entire distance?

Write your ideas below.

Thoughts About Running

_____

_____

_____

_____

_____

_____

_____

_____

II. Observe It

Pick a distance you can run and mark the starting and ending points. Using your feet, measure the distance by walking heel-to-toe between the two points. Count your steps and record the distance in "feet" below.

Now start a timer and run from the starting point to the ending point. Check the timer and see how long it takes.

Repeat this five times and record your answers below.

Times	
1	
2	
3	
4	
5	

Calculate your average speed. Add the above five times together and record the total time. Divide by 5 and record the time.

Total time _____ divided by 5 = _____ average speed

III. What Did You Discover?

❶ How fast did you go on your first run?

_____

❷ How fast did you go on your last run?

_____

❸ What was your average speed?

_____

❹ On your first run, did you run faster or slower than your average speed? Why?

_____

_____

_____

❺ On your last run, did you run faster or slower than your average speed? Why?

_____

_____

_____

❻ How far did you run? (How many feet did you measure?)

_____

_____

IV. Why?

In physics, the speed of an object is important to know. It is easy to measure the speed of an object that has linear motion—that is, an object that is moving from one point to the next in a straight line. You can measure your speed as you run by simply carrying a timer with you or having someone time you as you run. You can also find your average speed by repeating the run several times and averaging the times you record.

V. Just For Fun

How fast can your parent or friend run? Have them repeat the experiment using the same distance. Record their times. Calculate their average speed. Is their speed faster or slower than yours?

Times	
1	
2	
3	
4	
5	

Calculate the average speed. Add the above five times together and record the total time. Divide by 5 and record the time.

Total Time _____ **Divided By 5 =** _____ **Average Speed**

Experiment 8

Keep the Train on Its Tracks!

Introduction

This experiment is an exploration of how train wheels use rotational motion.

I. Think About It

❶ How does a train move on a track?

_____

_____

❷ What type of motion (linear or non-linear) are the train cars using? Why?

_____

_____

❸ What type of motion (linear or non-linear) are the wheels using? Why?

_____

_____

❹ How are train wheels shaped?

_____

_____

❺ What keeps the train wheels on the track?

_____

_____

_____

II. Observe It

❶ Take a cylinder that is 10-13 centimeters (4-5 inches) long and roll it on the ground. Note how far the cylinder rolls and whether it travels in a straight line. Try to trace the path of the cylinder with a piece of chalk. Record your observations.

Path of the Cylinder

❷ Take a plastic cup that has a mouth that is wider than the base. Roll it on the ground. Note how the plastic cup moves. Try to trace the path of the cup with a piece of chalk. Record your observations below.

Path of the Cup

❸ Take two long poles and tape them to the ground with about 5 centimeters (2 inches) between them.

❹ Roll the cylinder on the poles and observe how it moves. Note whether or not the cylinder stays on the poles. If it falls off, note how far it has moved. Record your observations below.

Movement of the Cylinder on the Poles

❺ Take two plastic cups and tape them together, end-to-end at the widest part (mouth).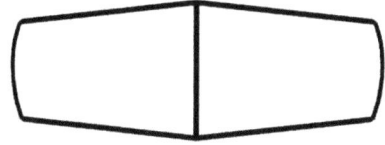

❻ Roll the taped plastic cups on the two poles. Note whether or not the taped plastic cups stay on the poles. If they fall off, note where and how far they rolled before falling off. Record your answers below.

Movement of the Taped Cups on the Poles

III. What Did You Discover?

❶ How far did the cylinder roll on the ground?

_____

_____

_____

_____

❷ Did the cylinder travel in a straight line? Why or why not?

_____

_____

_____

_____

❸ How far did the single plastic cup roll on the ground?

_____

_____

_____

_____

❹ Did the plastic cup travel in a straight line? Why or why not?

_____

_____

_____

_____

❺ How far did the cylinder travel on the taped poles?

_____

_____

_____

_____

❻ Was it easy for the cylinder to stay on the poles? Why or why not?

_____

_____

_____

_____

❼ How far did the taped plastic cups travel on the poles?

_____

_____

_____

_____

❽ Was it easy for the taped plastic cups to stay on the poles? Why or why not?

_____

_____

_____

_____

IV. Why?

If you look carefully at a wheel on a train, you can see that it is slightly tapered, meaning that the outside of the wheel is smaller than the inside of the wheel. By having tapered wheels, trains use the physics of rotational motion to stay on the tracks.

You observed that the two taped plastic cups didn't fall off the poles as they rolled, but that it was harder to keep the cylinder from sliding off. This is because as the plastic cups roll, the rotational motion of the larger middle (where the mouths are taped) is greater than the rotational motion of the smaller ends. This difference in rotational motion keeps the taped cups from falling off the poles.

Train wheels work in a similar way. As train wheels roll, the tapered wheels keep the train on the track!

V. Just For Fun

❶ What happens if you tape the small ends of the plastic cups together? Do you think the two cups will stay on the poles? Why or why not? On the following page, record what you think will happen.

❷ Try it and see what happens! Record your observations.

Ideas about what will happen if 2 cups with their small ends taped together are rolled on the poles

Observations about what happens when 2 cups with their small ends taped together are rolled on the poles

Experiment 9

Lemon Energy

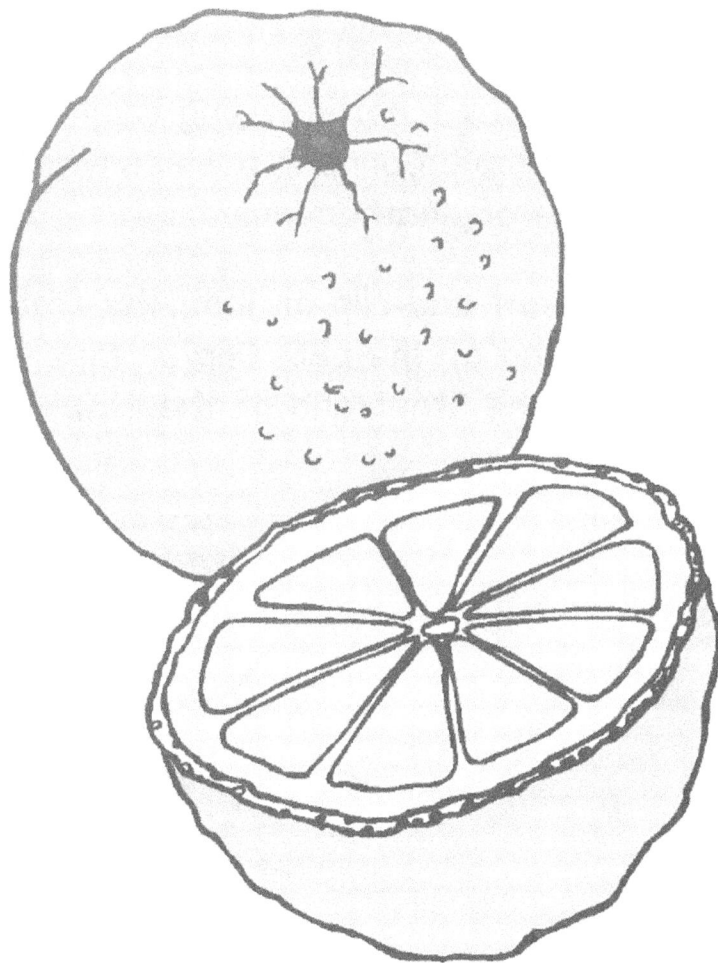

Introduction

Do you think lemons have stored chemical energy? Explore that idea in this experiment.

I. Observe It

❶ With the help of an adult, take three lemons and stick a copper penny in one end of each lemon and a zinc nail in the other end.

❷ Connect the three lemons together using copper wire and alligator clips or duct tape. Connect the penny side of one lemon to the zinc side of another lemon.

❸ Take the small LED light and clip or tape the free end of one of the wires to one end of the LED. Clip or tape the free end of the other wire to the other end of the LED.

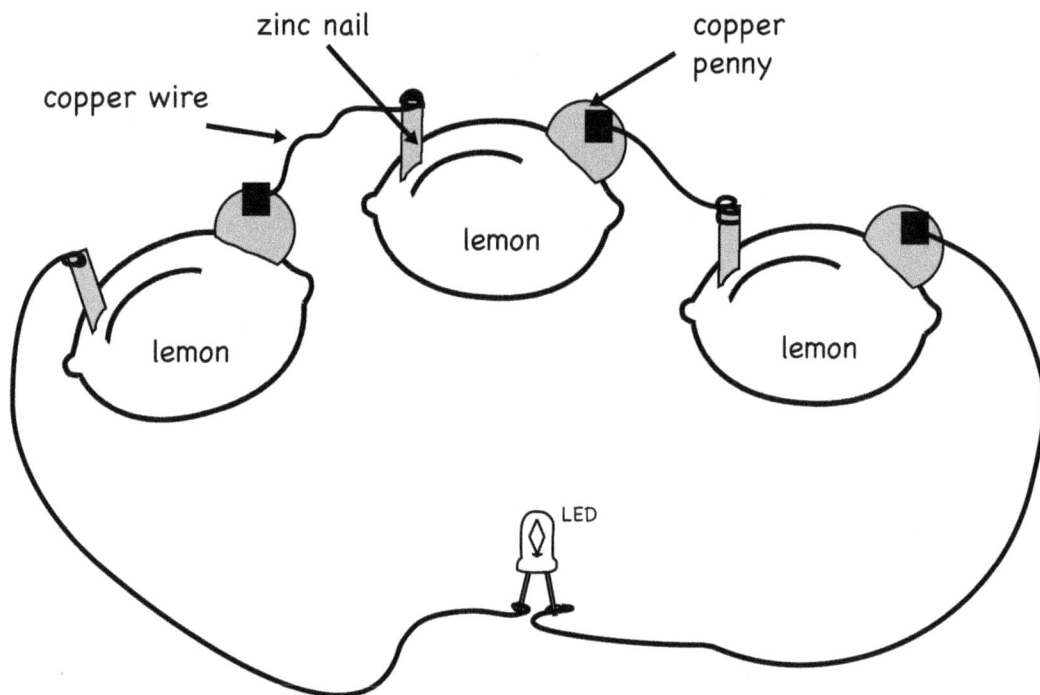

❹ Observe what happens when you attach the LED to the free ends of the wires.

❺ Record your observations below.

_____

_____

_____

_____

_____

_____

❻ Observe what happens to the LED when you disconnect one of the wires attached to a penny.

❼ Record your observations below.

_____

_____

_____

_____

_____

_____

❽ Reconnect the wire. Now observe what happens to the LED when you disconnect one of the wires attached to a zinc nail.

❾ Record your observations below.

_____

_____

_____

_____

_____

_____

Summarize Your Observations

Trial	What Happened?
All Wires Connected	
Penny Wire Disconnected	
Zinc Wire Disconnected	

II. Think About It

❶ Think about how you set up the lemon battery. Think about how you connected the wires, how you connected the LED to the lemons, and what happened then.

❷ Review what you learned about chemical energy and batteries in the textbook.

❸ Based on your observations, which of the following statements is true? Draw a circle around it.

The LED will light only when all the wires are connected.

The LED will light with the penny wire disconnected.

The LED will light when the zinc wire is disconnected.

❹ Think about any problems you might have had with your experiment. List them below.

_____

_____

_____

_____

_____

_____

_____

III. What Did You Discover?

❶ Were you able to get the LED to light up? Why or why not?

_____

_____

_____

_____

_____

_____

_____

❷ Did the LED light up when one or more of the wires was disconnected? Why or why not?

_____

_____

_____

_____

_____

_____

_____

IV. Why?

Before you did this experiment, did you know a lemon could be a battery? We normally don't think of lemons as batteries but as food. However, in this experiment you assembled an electric circuit using lemons as batteries. The lemons contain acid. The acid in the lemons reacts with the copper and zinc metals, and this creates a chemical reaction. This chemical reaction inside the lemons produces electricity that can be used to power a small LED. [The term LED stands for "light emitting diode" which is like a little light bulb that only requires a small amount of electricity to light up.]

Three lemons are used to light the LED. One lemon would not generate enough electricity to run the LED, so three lemons are needed. The lemons were connected to each other in such a way that the electricity of all the lemons could be added together. When all three lemons are working together, there is enough electricity generated by the chemical reactions of the combined lemons to light up the LED.

When one or more of the wires is disconnected, the LED stops working. By disconnecting the wires, the battery energy in each lemon can no longer be added to that of the other lemons, the electricity cannot flow to the LED, and so the LED stops working. The lemons are a form of stored chemical energy, and once they are connected to each other in the right way, they can generate enough electricity to power a small LED.

V. Just For Fun

Disconnect one of the wires from the LED. With the fingers of one of your hands, hold the wire connected to the lemons. With the fingers of your other hand, hold the metal end on the disconnected side of the LED.

What happens?

_____

_____

_____

_____

_____

_____

Experiment 10

Sticky Balloons

Introduction

With this experiment, see if you can find out how much charge a balloon has.

I. Observe It

❶ Take a rubber balloon and blow it up with air. Tie the end closed, then place the balloon on a wall. Observe whether it sticks to the wall.

❷ Without popping the balloon, rub it in your hair.

❸ Carefully pull the balloon away from your hair and observe whether your hair sticks to the balloon. If your hair sticks to the balloon, continue to Step **❹**. If your hair does not stick to the balloon, rub the balloon in your hair again.

❹ Test how sticky the balloon is by placing it on a wall. Observe whether the balloon sticks or falls off the wall.

❺ Record your observations in the space provided on the next page. Use the following questions.

➤ Does the balloon stick?

➤ How long does the balloon stick? 1 second? 2 seconds? 10 seconds? Longer than a minute?

➤ Does the balloon move around or stay still?

➤ What happens if you blow gently on the balloon? Does it stay stuck or does it fall off?

Hair

_____

_____

_____

_____

_____

_____

_____

❻ Repeat Steps ❷-❺. This time, instead of rubbing the balloon in your hair, rub it on different materials, such as wool or cotton clothing, metal or wood surfaces. In the following boxes record your observations for each object.

_____

_____

_____

_____

_____

_____

_____

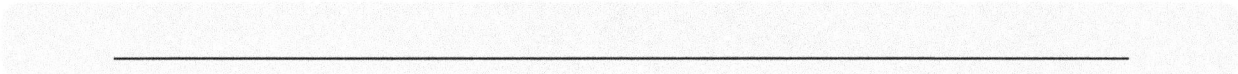

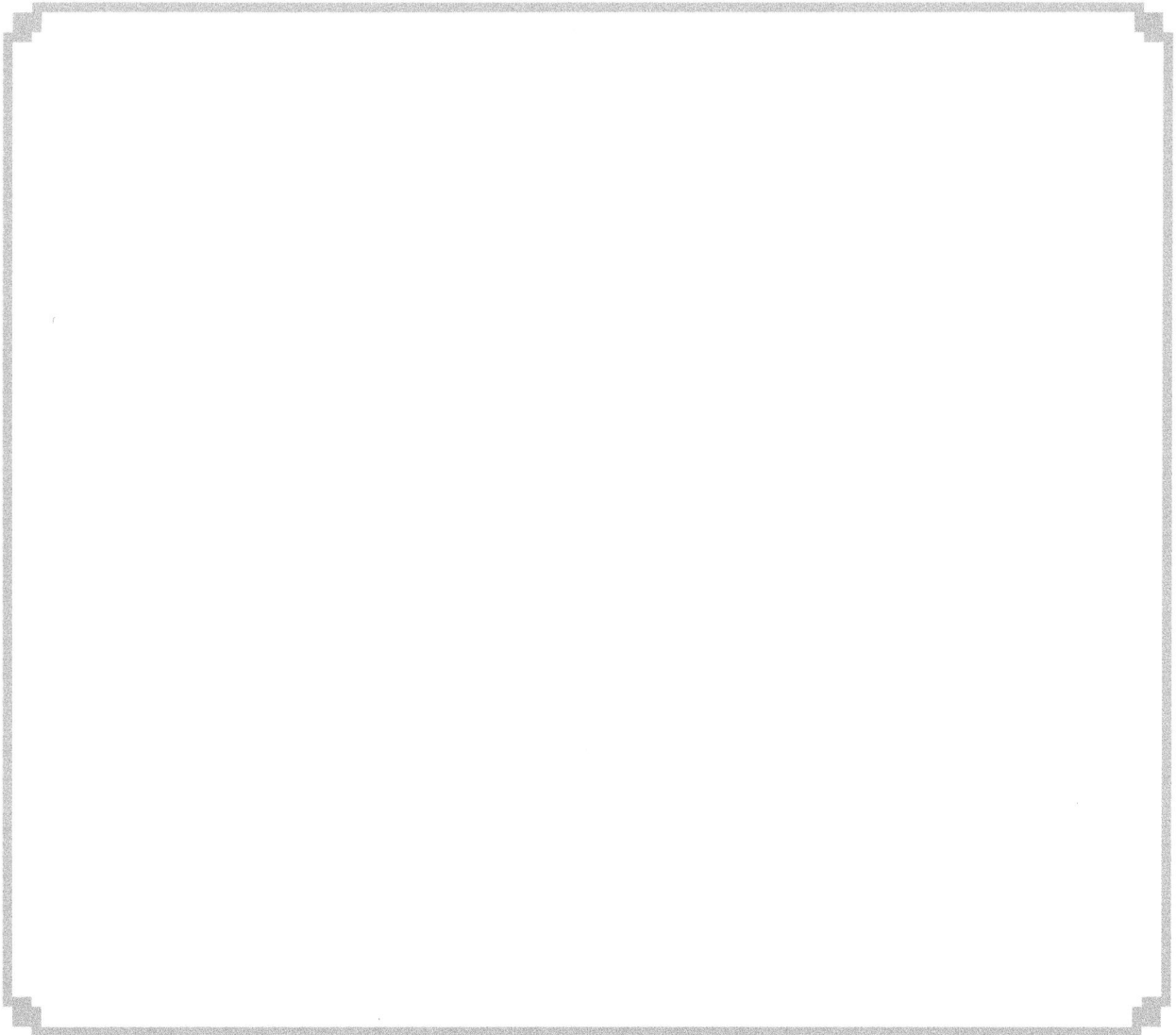

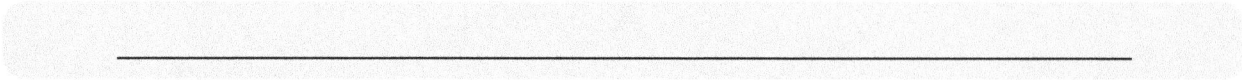

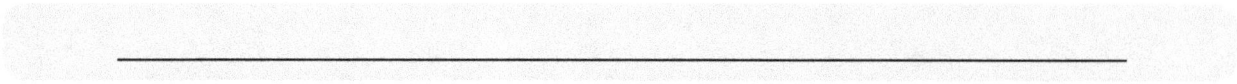

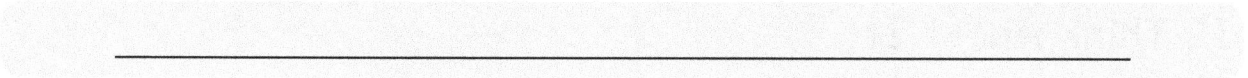

II. Think About It

❶ Think about the balloon and the different materials you used to charge the balloon.

❷ Review what you learned in the textbook about electrons, charges, and force.

❸ Create a chart below that lists the materials or surfaces you used in your experiment. Begin with those materials or surfaces that created the most charge, and end with those that created the least charge.

Most Charge

↓

Least Charge

III. What Did You Discover?

❶ Were you able to get the balloon to carry a charge? Why or why not?

_____

_____

_____

_____

❷ Did some materials charge the balloon more than other materials, or were they all the same?

_____

_____

_____

_____

❸ Did attaching the balloon to the wall allow you to tell how much charge the balloon had gained? Why or why not?

_____

_____

_____

_____

IV. Why?

In this experiment you explored how a balloon can pick up electrons from other objects. When a balloon picks up electrons from another object, the balloon becomes charged. The very first time you placed the balloon on the wall, before you rubbed it in your hair, the balloon likely fell off. Why? It fell off because the balloon did not carry any additional charges.

When you rubbed the balloon in your hair, the balloon picked up electrons from your hair. The electrons are negatively charged, so the balloon became negatively charged. A negatively charged balloon will stick to surfaces that are slightly positively charged. If the balloon stuck to the wall, then the wall was slightly positively charged.

You could test how many electrons the balloon picked up by observing how easily the balloon would stick to a wall after it was charged. If the balloon stuck a lot, then there were lots of electrons picked up. If the balloon only stuck a little, then there were fewer electrons.

It is possible that it was difficult to get the balloon to become charged no matter what material or surface was used. If you discovered this, that's OK. Check the humidity in your area. If it was humid when you did the experiment, the electrons could not stay very long on the balloon.

V. Just For Fun

Take two balloons and tie a piece of string to the end of each balloon. Tie the other ends of the strings together and hang the balloons from a doorway or shower rod. What happens?

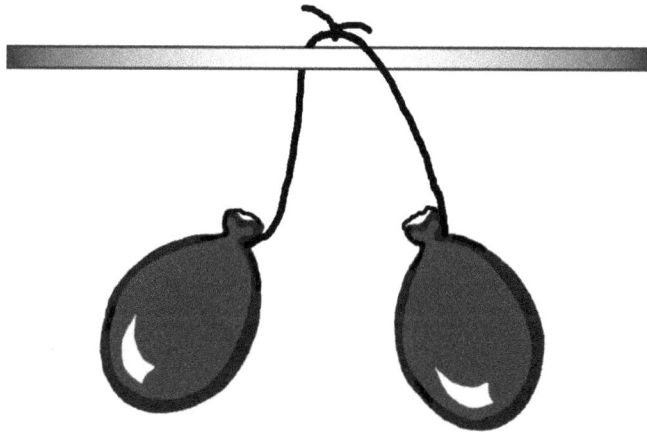

Take the balloons and rub them both in your hair. Let the balloons go and allow them to float back together. What happens?

Record your observations below.

Before Rubbing	**After Rubbing**

Experiment 11

Moving Electrons

Introduction

Discover whether using different materials can affect the flow of electrons in a moving electric current.

I. Observe It

❶ With the help of an adult, set up the lemon battery from the *Lemon Energy* experiment. Make sure the LED is illuminated.

❷ With the help of an adult, take one of the wires between two lemons and cut it. Observe what happens to the LED. Write or draw your observations below.

Wires Apart

_____

_____

_____

_____

❸ Reconnect the ends of the wire. Observe what happens to the LED. Write or draw your observations below.

Wires Connected

_____

_____

_____

_____

_____

❹ Disconnect the ends of the cut wire and place a piece of Styrofoam between them. Observe what happens to the LED. Write or draw your observations below.

Styrofoam

_____

_____

_____

_____

_____

❺ Remove the Styrofoam from between the wires. Reconnect the wires by twisting them together. Observe what happens to the LED. Write or draw your observations below.

Wires Connected

⓺ Repeat Steps **⓸**-**⓹** using the different materials on the following pages. Test all of the materials listed. Record your observations in the space provided.

Plastic Block

_____

_____

_____

_____

_____

_____

Cotton Ball

_____

_____

_____

_____

_____

_____

Nickel Coin

_____

_____

_____

_____

_____

_____

Metal Paperclip

_____

_____

_____

_____

_____

_____

Plastic Paperclip

_____

_____

_____

_____

_____

_____

❼ Summarize your results below. Write "ON" or "OFF" in the LED column for each of the items listed.

Item	LED
Start—wires connected	
Wires apart	
Wires connected	
Styrofoam	
Wires connected	
Plastic block	
Cotton ball	
Nickel coin	
Metal paperclip	
Plastic paperclip	

II. Think About It

❶ Think about the LED and the different materials you placed between the wires.

❷ Review what you learned in your *Student Textbook* about moving electric charges.

❸ Use the chart below to organize the test results for the various materials. Place those items that illuminated the LED in one column and those items that did not illuminate the LED in the other column.

LED "ON"	LED "OFF"

III. What Did You Discover?

❶ Which items illuminated the LED?

_____

_____

_____

❷ Which items did not illuminate the LED?

_____

_____

_____

❸ Were the items that illuminated the LED metals?

_____

_____

_____

❹ Were the items that did not illuminate the LED non-metals?

_____

_____

_____

❺ Why do you think metals illuminated the LED and non-metals did not?

_____

_____

_____

IV. Why?

In this experiment you explored how electrons moved (or didn't move) through different materials. When the lemons in the battery are connected with metal wires, the electrons can flow freely through the wires and light up the LED. When the metal wire is cut, the electrons are stopped from flowing through to the LED. When the metal wires are reconnected, the electrons flow through the wire again to light up the LED.

You discovered that some materials will allow electrons to flow through them, and some materials won't. Electrons easily flow through most metals. Metals and any other materials that allow electrons to flow through them are called *conductors*. Electrons do not flow through most plastics (Styrofoam, plastic blocks, or plastic paperclips), cotton balls, and other similar materials. These materials are called *insulators*.

There are also materials that are reluctant to allow electrons to flow through them. These materials are called *resistors*. Resistors will allow a few electrons to flow through them, but not all electrons. Resistors are used in electronic circuits to control the amount of electron flow.

V. Just For Fun

Take the two ends of the wire that was cut. Place both ends in a glass of water without letting them touch each other. Observe what happens to the LED.

Now add 15 ml (1 tablespoon) of salt to the water. Stir the salt until it has completely dissolved. Again place the wire ends in the glass and observe what happens to the LED.

Record your observations below.

Ends of Wire in Water

Ends of Wire in Saltwater

Experiment 12

Magnet Poles

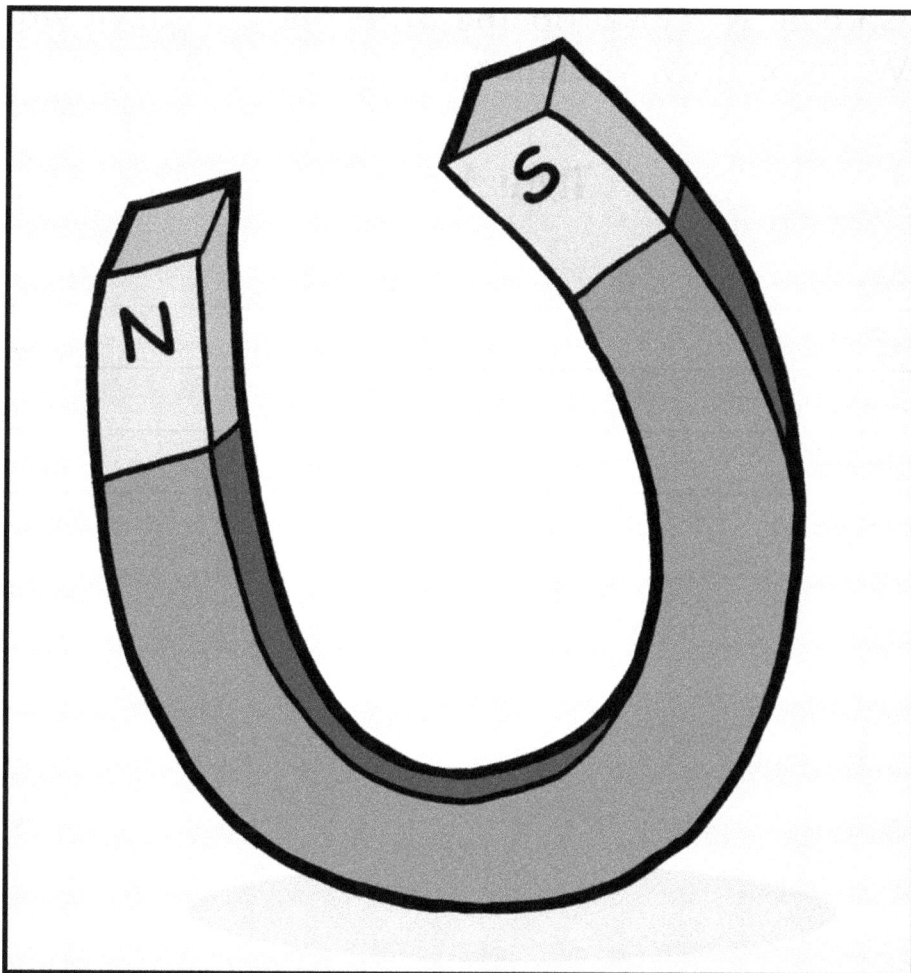

Introduction

Explore how magnet poles behave!

I. Observe It

❶ Take two magnets and place them on a table several inches apart with each "**N**" facing the other.

❷ Gently push one "**N**" closer to the other "**N**." Observe what happens. Write or draw your observations below.

Trial 1

_____

_____

_____

_____

❸ Place the two magnets on the table several inches apart. Reverse the direction of one of the magnets so that the "**S**" of one is facing the "**N**" of the other.

❹ Gently push the "**N**" closer to the "**S**." Observe what happens. Write or draw your observations below.

Trial 2

_____

_____

_____

_____

❺ Repeat Steps ❶-❹ several times. See how close you can bring the two magnets together before something changes.

Write or draw your observations in the spaces provided.

Trial 3

_____

_____

_____

_____

_____

_____

_____

Trial 4

_____
_____
_____
_____
_____
_____
_____

Trial 5

_____

_____

_____

_____

_____

_____

_____

Trial 6

_____

_____

_____

_____

_____

_____

Trial 7

_____

_____

_____

_____

_____

_____

_____

Trial 8

_____

_____

_____

_____

_____

_____

_____

Trial 9

Trial 10

_____

_____

_____

_____

_____

_____

_____

❻ Summarize your results in the chart.

➡ Mark the trials **N-N** or **N-S**.

➡ Mark the trials where the magnets came together.

➡ Mark the trials where the magnets pushed apart.

Trial	N-N or N-S	Together or Apart
Trial 1		
Trial 2		
Trial 3		
Trial 4		
Trial 5		
Trial 6		
Trial 7		
Trial 8		
Trial 9		
Trial 10		

II. Think About It

❶ Think about the magnets and the different ways you pushed the magnets together.

❷ Review what you learned in your *Student Textbook* about magnets and magnetic poles.

❸ Look at the results you gathered in the previous section. Cover up the second column of the table and look only at the first and third columns. Without looking at the second column, write down those trials where the poles were the "same" and those trials where the poles were "opposite."

Trial	Same or Opposite
Trial 1	
Trial 2	
Trial 3	
Trial 4	
Trial 5	
Trial 6	
Trial 7	
Trial 8	
Trial 9	
Trial 10	

❹ Uncover the second column from the previous page. Do your answers match the second column of the previous table?

III. What Did You Discover?

❶ What happened when you pushed the two "**N**" ends of the magnets together?

_____

_____

_____

❷ What happened when you reversed one of the magnets and pushed them together again?

_____

_____

_____

❸ If you had two magnets where the poles were not labeled "**N**" and "**S**," could you guess which poles were the same and which were opposite? Why or why not?

_____

_____

_____

_____

_____

IV. Why?

In this experiment you explored magnetic poles. Magnets have two poles, one called "north" and one called "south." When the two opposite poles (north and south) come together, they attract each other, and the magnets will snap together. When two of the same pole (north and north, or south and south) come together, they repel each other, and the magnets will move away from each other.

Even though you may not know which pole is "north" and which pole is "south," by "playing" with the magnets (by reversing one of the magnets on the table and then switching it back), you can explore how the different poles react to each other. Reversing the magnet several times gives you information about when the magnets are coming together and when they are moving apart. As you observe the magnets coming together or moving apart, you are observing the effects of the different poles and can tell which poles are the same and which are opposite.

Scientists have to "play" with things around them to figure out what is happening. Scientists do different trials, just like you did, to find out what happens when some part of an experiment is changed. By "playing" with their experiments, scientists make observations they might have missed if they did the experiment only one way.

V. Just For Fun

Take a magnet and find out which surfaces in your house the magnet will attract or not attract.

Record your observations below.

Surface	Attract or Not Attract

More REAL SCIENCE-4-KIDS Books
by Rebecca W. Keller, PhD

Building Blocks Series yearlong study program — each Student Textbook has accompanying Laboratory Notebook, Teacher's Manual, Lesson Plan, Study Notebook, Quizzes, and Graphics Package

Exploring the Building Blocks of Science Book K (Activity Book)
Exploring the Building Blocks of Science Book 1
Exploring the Building Blocks of Science Book 2
Exploring the Building Blocks of Science Book 3
Exploring the Building Blocks of Science Book 4
Exploring the Building Blocks of Science Book 5
Exploring the Building Blocks of Science Book 6
Exploring the Building Blocks of Science Book 7
Exploring the Building Blocks of Science Book 8

Focus Series unit study program — each title has a Student Textbook with accompanying Laboratory Notebook, Teacher's Manual, Lesson Plan, Study Notebook, Quizzes, and Graphics Package

Focus On Elementary Chemistry
Focus On Elementary Biology
Focus On Elementary Physics
Focus On Elementary Geology
Focus On Elementary Astronomy

Focus On Middle School Chemistry
Focus On Middle School Biology
Focus On Middle School Physics
Focus On Middle School Geology
Focus On Middle School Astronomy

Focus On High School Chemistry

Super Simple Science Experiments

21 Super Simple Chemistry Experiments
21 Super Simple Biology Experiments
21 Super Simple Physics Experiments
21 Super Simple Geology Experiments
21 Super Simple Astronomy Experiments
101 Super Simple Science Experiments

Note: A few titles may still be in production.

Gravitas Publications Inc.
www.gravitaspublications.com
www.realscience4kids.com

GRAVITAS
PUBLICATIONS

www.ingramcontent.com/pod-product-compliance
Lightning Source LLC
Chambersburg PA
CBHW080558220326
41599CB00032B/6526